Pollinators

BEETLES

Martha London

DiscoverRoo
An Imprint of Pop!
popbooksonline.com

abdobooks.com

Published by Pop!, a division of ABDO, PO Box 398166, Minneapolis, Minnesota 55439. Copyright © 2020 by POP, LLC. International copyrights reserved in all countries. No part of this book may be reproduced in any form without written permission from the publisher. Pop!™ is a trademark and logo of POP, LLC.

Printed in the United States of America, North Mankato, Minnesota.

102019
012019

THIS BOOK CONTAINS RECYCLED MATERIALS

Cover Photo: Howard Stapleton/Alamy
Interior Photos: Howard Stapleton/Alamy, 1; Shutterstock Images, 5, 7 (beetle), 11, 19, 21, 23 (top), 25, 28, 29, 30; iStockphoto, 6, 7 (flower), 8–9, 12, 13, 14, 20, 22 (bottom), 23, 26–27, 31; Matthias Lenke/Science Source, 15; Clement Philippe/Arterra Picture Library/Alamy, 17; Gilbert S. Grant/Science Source, 18

Editor: Connor Stratton
Series Designer: Jake Slavik

Library of Congress Control Number: 2019942633

Publisher's Cataloging-in-Publication Data
Names: London, Martha, author.
Title: Beetles / by Martha London
Description: Minneapolis, Minnesota : Pop!, 2020 | Series: Pollinators | Includes online resources and index.
Identifiers: ISBN 9781532165948 (lib. bdg.) | ISBN 9781532167263 (ebook)
Subjects: LCSH: Pollinators--Juvenile literature. | Beetles--Juvenile literature. | Beetles--Behavior--Juvenile literature. | Pollination by insects--Juvenile literature. | Insects--Juvenile literature.
Classification: DDC 595.76--dc23

WELCOME TO
DiscoverRoo!

Pop open this book and you'll find QR codes loaded with information, so you can learn even more!

Scan this code* and others like it while you read, or visit the website below to make this book pop!

popbooksonline.com/beetles

*Scanning QR codes requires a web-enabled smart device with a QR code reader app and a camera.

TABLE OF CONTENTS

CHAPTER 1
Dropping Pollen. 4

CHAPTER 2
Beetle Bodies. 10

CHAPTER 3
Environments 16

CHAPTER 4
Beetles in the World. 24

Making Connections 30
Glossary. 31
Index 32
Online Resources 32

CHAPTER 1
DROPPING POLLEN

The summer sun is bright and warm. A beetle moves across a pond. It climbs onto the petal of a lotus flower. The beetle eats the flower's **pollen**. It eats some petals too.

WATCH A VIDEO HERE!

A stag beetle visits a lotus flower.

Beetles are the main pollinators of lotus flowers.

After eating, the beetle moves to another lotus flower. The beetle leaves droppings inside this new flower. The droppings have pollen inside them. This pollen gets into the second plant.

BEETLE POLLINATION

A beetle eats pollen from a flower. The beetle flies to another flower. It leaves droppings in the second flower. Those droppings have pollen from the first flower.

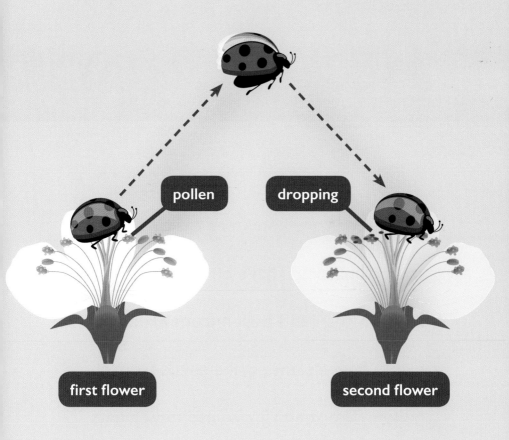

Beetles are known as mess and soil pollinators. That's because their droppings spread pollen from plant to plant. These plants make seeds from the pollen. The seeds grow into new plants.

Goldenrod is one type of flower that beetles pollinate.

Other animals act as pollinators too.

But some plants need beetles. Without beetles, these plants could not survive.

 **Beetles pollinate 88 percent of all the flowering plants in the world.**

CHAPTER 2
BEETLE BODIES

More than 360,000 **species** of beetles exist in the world. Beetles are insects. Like all insects, they are **invertebrates**. They have hard shells and soft bodies.

COMPLETE AN ACTIVITY HERE!

The rhinoceros beetle can pick up more than 850 times its weight. That is like a person picking up nine elephants.

DID YOU KNOW? Beetles make up approximately 40 percent of all types of insects.

Each beetle has two pairs of wings. One pair is for flying. These wings are thin and clear. They can easily break.

hard, outer wing

soft, inner wing

Many beetles fly with their legs stretched forward. They use their legs to help them turn.

A second pair of wings protect the thin ones. Those wings are hard. When beetles fly, the hard wings spread out to the sides. The flying wings flap.

Most insects do not chew their food. But beetles' mouthparts allow them to chew **pollen**. They use jaw parts

Male stag beetles use their huge jaws to fight other male stag beetles.

to bite. Parts known as palps act like fingers. They help beetles get food into their mouths.

CHAPTER 3
ENVIRONMENTS

Beetles are found all over the world. They are able to live in every **climate**. Beetles can also live underground or underwater. They are very good at **adapting** to their **environment**.

LEARN MORE HERE!

air bubble

Many diving beetles carry air bubbles with them underwater. They use these bubbles to breathe.

For example, one type of beetle lives among trees that are green all year. This beetle is bright green. It can easily hide among the leaves. As the beetle hides, it eats leaves.

A glorious scarab beetle blends in with a juniper tree.

When a tiger beetle runs, its brain cannot work fast enough to make sense of what it sees. The beetle goes blind and stops running. Then its brain catches up.

Tiger beetles are the fastest insects on the planet. They can cover 120 body lengths per second! Humans can only cover five lengths.

Adult beetles' environments are different from young beetles' environments. For instance, many

Ladybug beetle eggs rest on a leaf.

Bark beetles lay eggs inside the wood of trees.

beetles lay eggs inside flowers or fruit. When young beetles hatch, they eat their way out. As adults, beetles spread out. They look for other plants to eat.

LIFE CYCLE OF A BEETLE

Female beetles lay eggs. Some **species** lay hundreds at a time.

Eggs hatch into beetle larvae. Larvae are often shaped like worms.

As larvae grow, they shed their skin. Then they enter cocoons.

Some beetles live for just 13 weeks. Others live for up to 12 years.

Adult beetles emerge from their cocoons.

CHAPTER 4
BEETLES IN THE WORLD

Certain kinds of beetles can be harmful. For instance, some beetles eat crops that people depend on for food. Other beetles dig into trees. They lay their eggs inside the trees. Then the larvae

LEARN MORE HERE!

Colorado potato beetles eat through potato plants.

eat through the trees. As a result, those trees often die.

Magnolias are a type of flowering tree.

But most beetles are helpful. In fact, beetles were one of the first insect pollinators on Earth. Certain plant **species**, such as magnolias, started

THE FIRST POLLINATORS

Beetles have existed for at least 200 million years. Dinosaurs still walked the planet during that time. Flowering plants developed at least 140 million years ago. Scientists believe these first flowers looked like magnolias. Their petals were tightly pressed together. Beetles crawled through them to find **pollen**.

growing long before bees even existed. Beetles still pollinate these types of plants today.

Some types of beetles also act as **scavengers**. These beetles eat dead animals. Some also eat dead plants. They help break down dead material. And their droppings add nutrients to soil. The nutrients help plants grow. Without beetles, many plants would not survive.

One type of scavenger is the burying beetle.

The passalid beetle helps break down dead wood.

MAKING CONNECTIONS

TEXT-TO-SELF

Have you seen beetles before? If so, what kinds? If not, where might you find them?

TEXT-TO-TEXT

Have you read books about other insects? What do they have in common with beetles? How are they different?

TEXT-TO-WORLD

Beetles can be both helpful and harmful. What other animals both help and hurt the world around them?

GLOSSARY

adapt – to develop traits that make it easier to survive in an environment.

environment – the natural surroundings where plants and animals live.

invertebrate – a type of animal without a spine.

pollen – fine, dust-like stuff that flowers create and use to reproduce.

scavenger – an animal that eats creatures that are already dead.

species – a group of animals or plants of the same kind that can reproduce.

INDEX

cocoons, 23

droppings, 6, 7, 8, 28

eggs, 21, 22, 24

invertebrates, 10

jaw parts, 14

larvae, 22–23, 24

lotus flowers, 4, 6

magnolias, 26, 27

palps, 15

scavengers, 28

tiger beetles, 19

wings, 12–13

ONLINE RESOURCES
popbooksonline.com

Scan this code* and others like it while you read, or visit the website below to make this book pop!

popbooksonline.com/beetles

*Scanning QR codes requires a web-enabled smart device with a QR code reader app and a camera.